Area at

MW01594951

by Barbara Andrews

Table of Contents

Introduction

We see shapes in the park. All closed shapes have **area**. We measure to find the area.

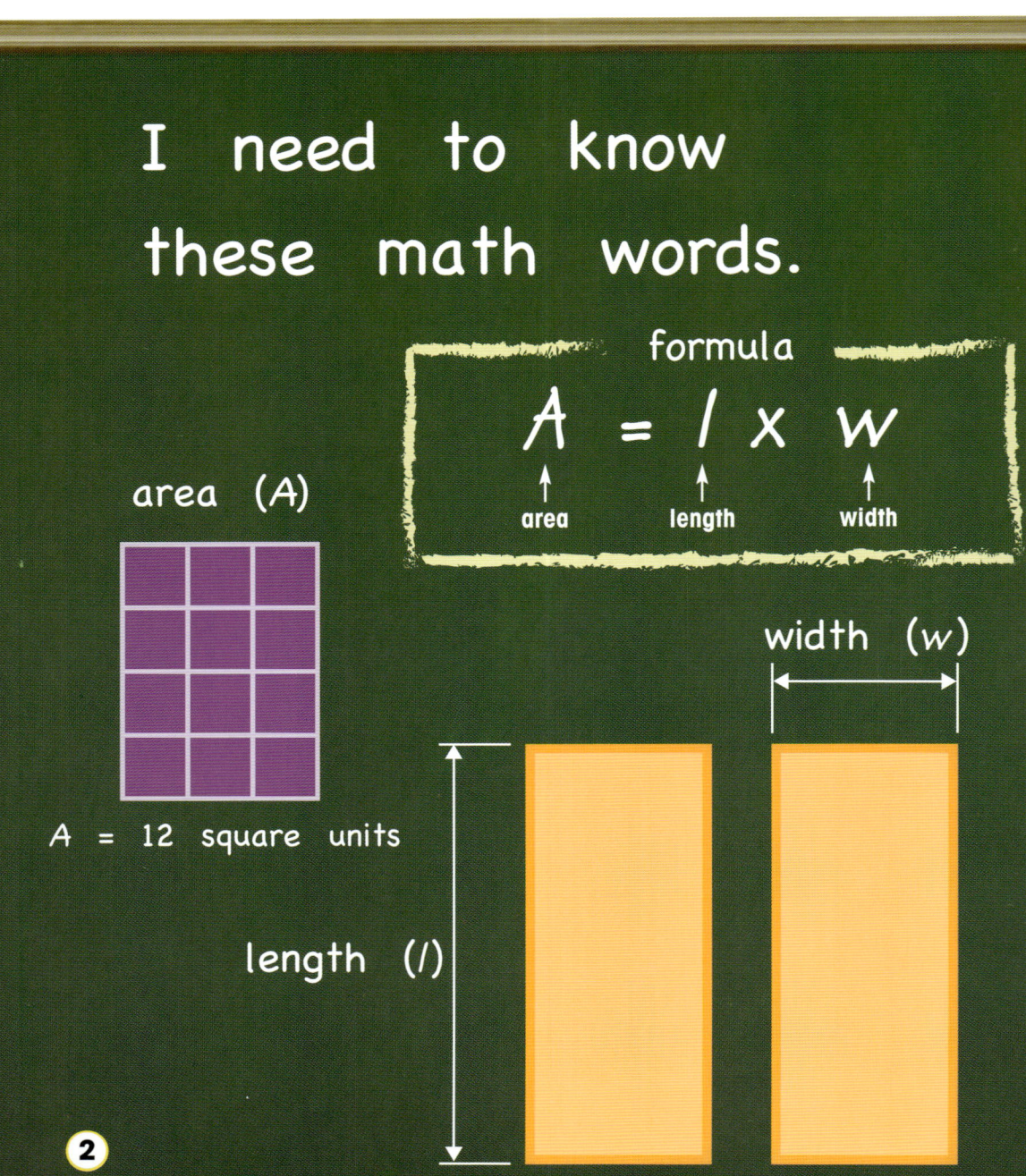

I need to know
these math words.

formula

$$A = l \times w$$

↑ area   ↑ length   ↑ width

area (A)

A = 12 square units

width (w)

length (l)

rectangle

ruler

square units

square unit

square

What Is Area?

Area is a measurement. We use **square units** to measure area. How many square units cover this shape?

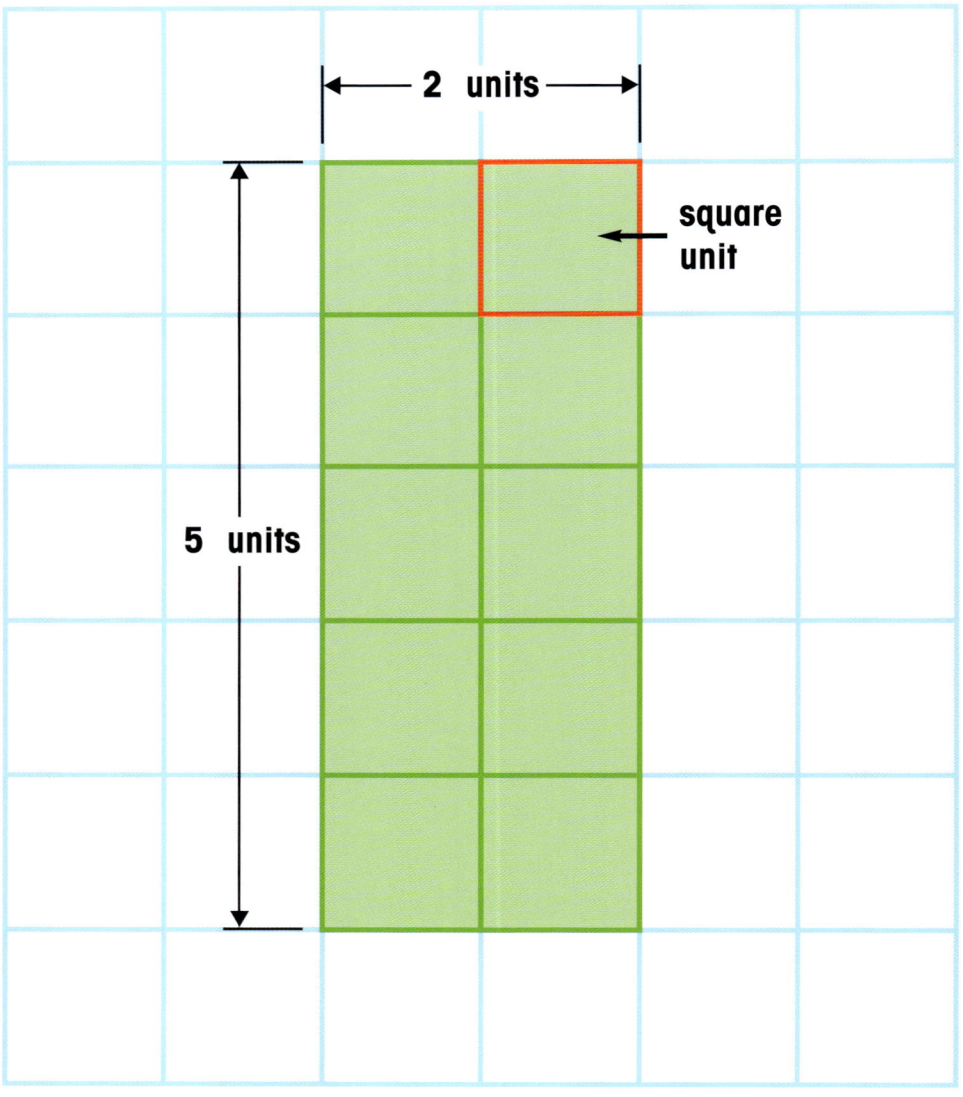

▲ **Ten square units cover this shape.**

Every closed shape has area. Every closed shape has square units. This **rectangle** has square units.

1	2	3	4	5
6	7	8	9	10
11	12	13	14	15
16	17	18	19	20

▲ **Count the number of square units.**

Solve This

Find the area of this shape.
Count square units to find the area.

Answer: 9 square units

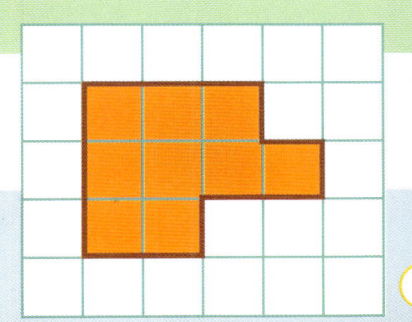

We can multiply to find the area. We multiply to find a product. We can find the area of this rectangle.

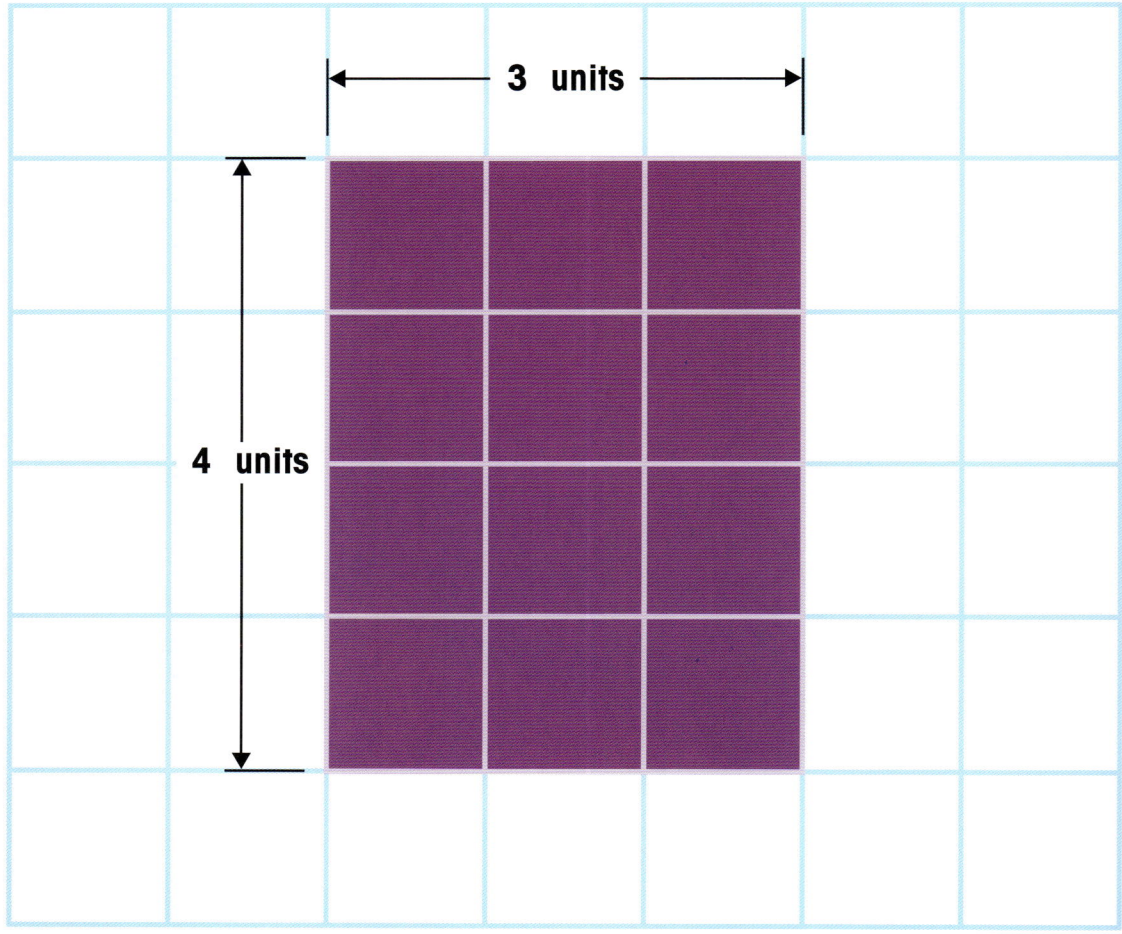

3   x   4   =   12

▲ The area is twelve square units.

This **square** has an area. Each side is three units long. The area is nine square units.

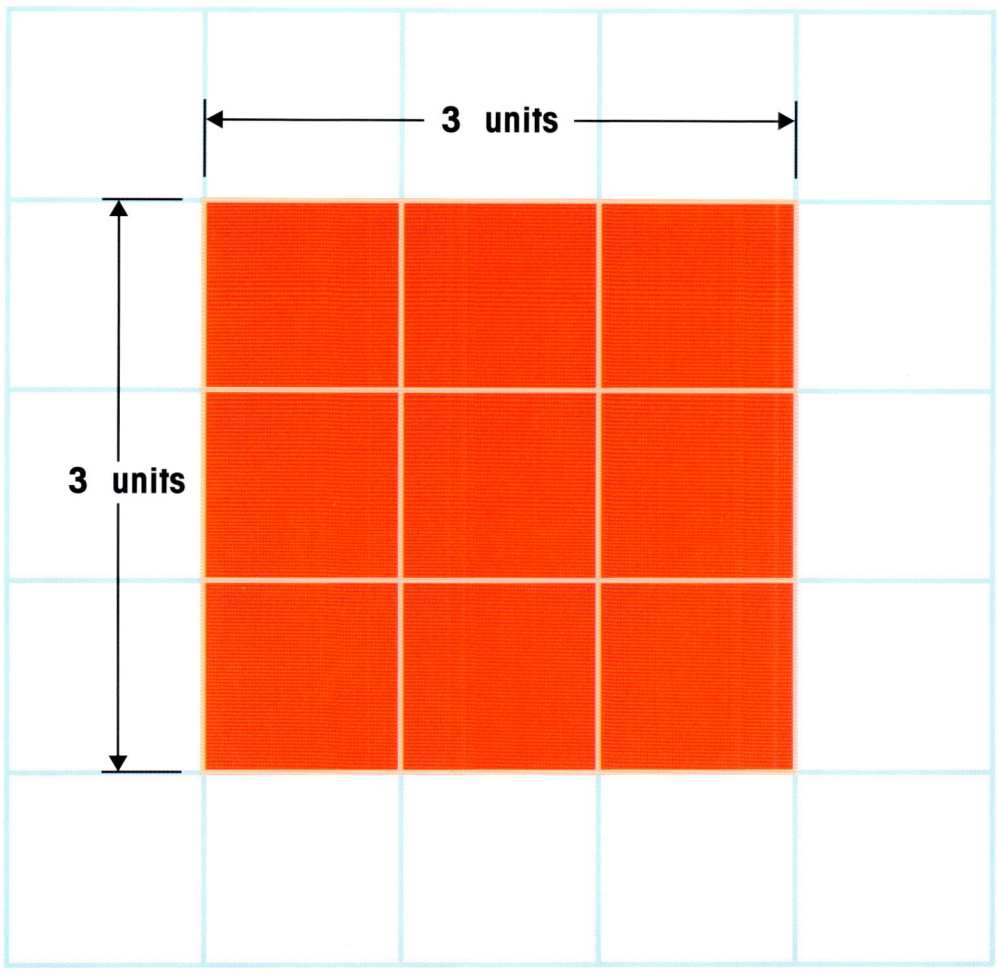

3 units

3 units

3 x 3 = 9

▲ This square has nine square units.

7

How Do We Find the Area?

We can find the area of a rectangle. We use a **ruler**. We measure the sides.

▲ We use a ruler to measure.

First, we measure the **length**. Then we measure the **width**.

◀ We measure the length.

▲ We measure the width.

Then we use a **formula**. This formula helps us find the area. We multiply the length by the width.

$$A = l \times w$$

Area is equal to length multiplied by width.

▲ **The formula tells us to multiply.**

▲ **The area is seventy-two square feet.**

The product is the area. We find the area of the rectangle.

How to Find the Area of a Rectangle

1. Measure the length.
2. Measure the width.
3. Multiply the length by the width.

Solve This

Find the area of this rectangle.
Use the formula on page 10.

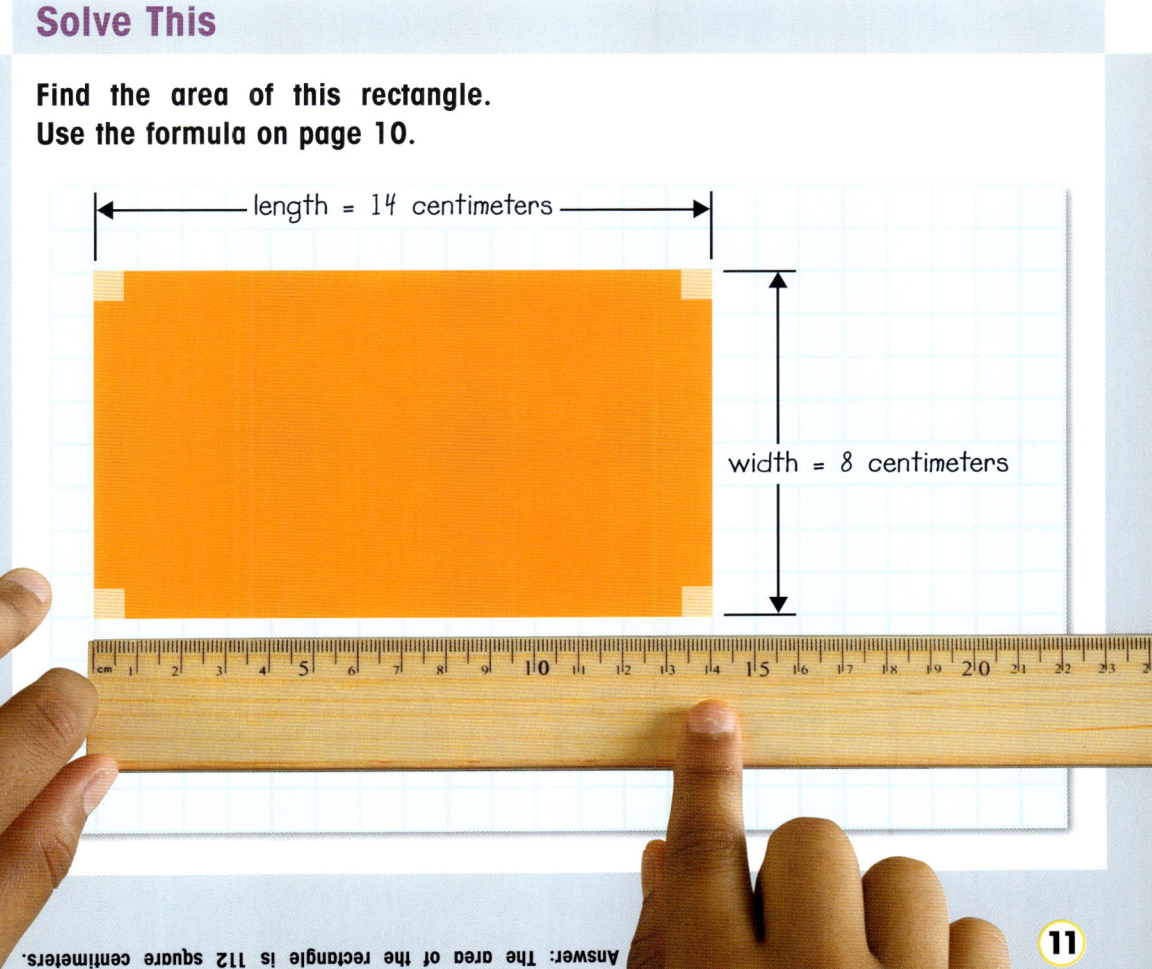

length = 14 centimeters

width = 8 centimeters

Answer: The area of the rectangle is 112 square centimeters.

What Has Area in the Park?

This garden has an area. The length is nine feet. The width is nine feet.

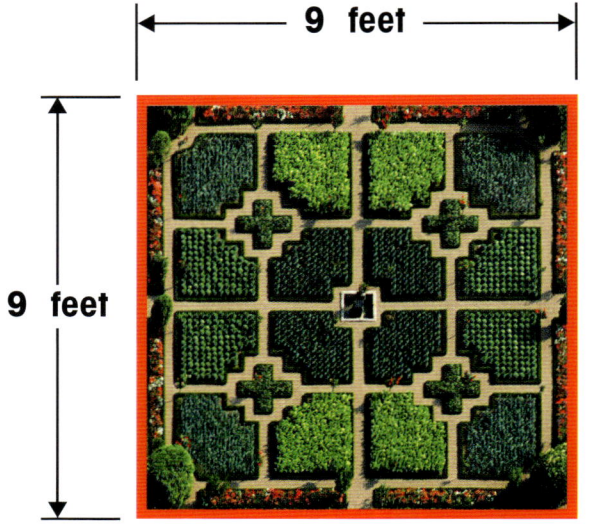

9 feet

9 feet

$$A = l \times w$$
$$A = 9 \times 9$$
$$A = 81 \text{ square feet}$$

▲ The area is eighty-one square feet.

Figure It Out

Look at the garden.
What shape is the garden?
How do you know?

This picnic blanket has an area. The length is sixty inches. The width is thirty-five inches.

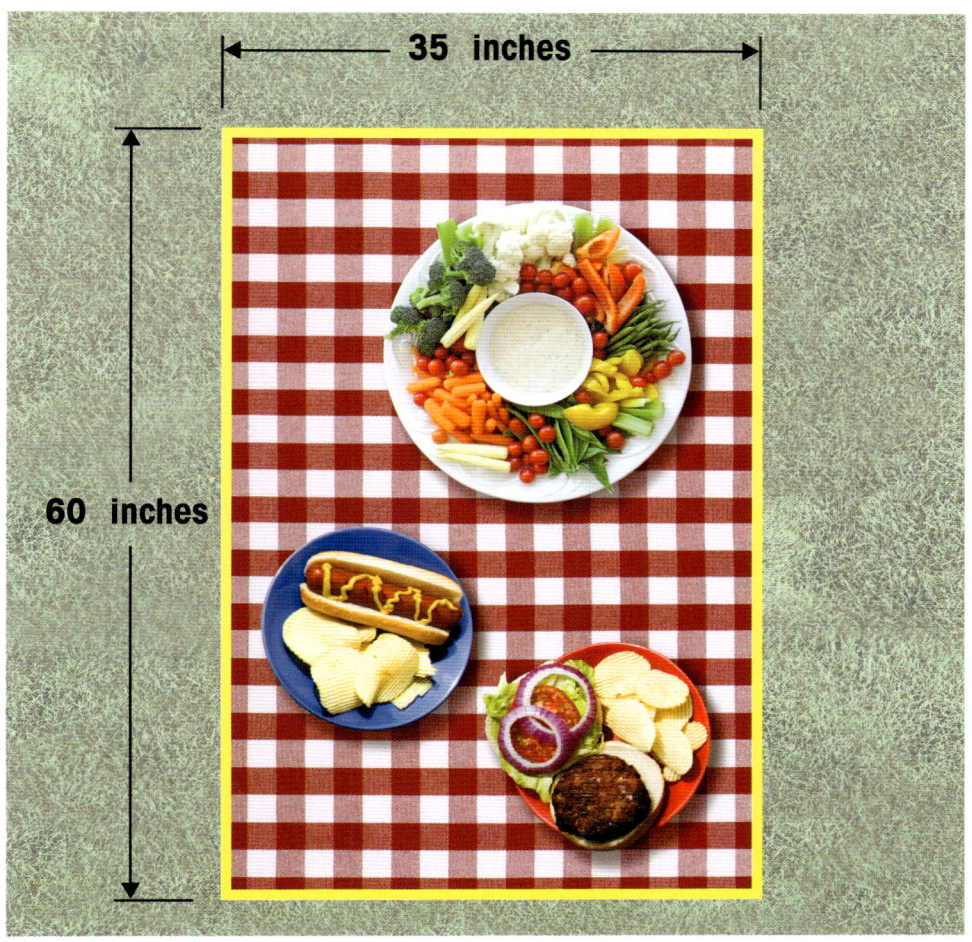

35 inches

60 inches

$$A = l \times w$$
$$A = 60 \times 35$$
$$A = 2,100 \text{ square inches}$$

▲ The area is two thousand one hundred square inches.

This basketball court has an area. The length is forty-seven feet. The width is twenty-five feet.

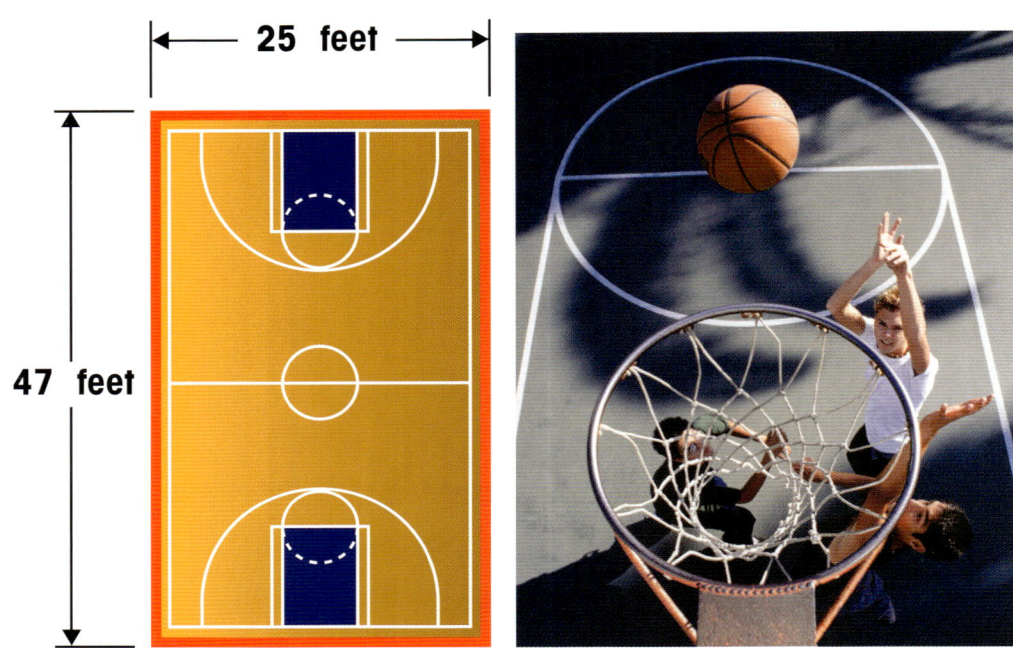

25 feet

47 feet

$$A = l \times w$$
$$A = 47 \times 25$$
$$A = 1,175 \text{ square feet}$$

▲ **The area is one thousand one hundred seventy-five square feet.**

Solve This

This basketball court has an area. Find the area.

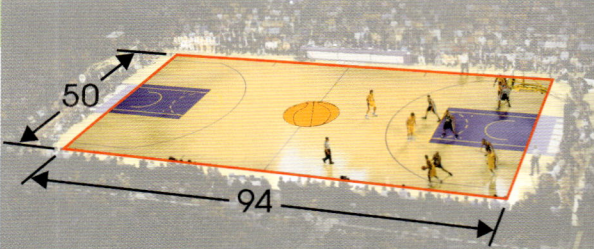

50

94

Answer: 4,700 square feet

This grill has an area. The length is sixty centimeters. The width is fifty centimeters.

50 centimeters

60 centimeters

$A = l \times w$
$A = 60 \times 50$
$A = 3,000$ square centimeters

▲ The area is three thousand square centimeters.

This tennis court has an area. The length is seventy-eight feet. The width is thirty-six feet.

$$A = l \times w$$
$$A = 78 \times 36$$
$$A = 2,808 \text{ square feet}$$

▲ The area is two thousand eight hundred eight square feet.

This playground has an area. The length is thirty meters. The width is twenty meters.

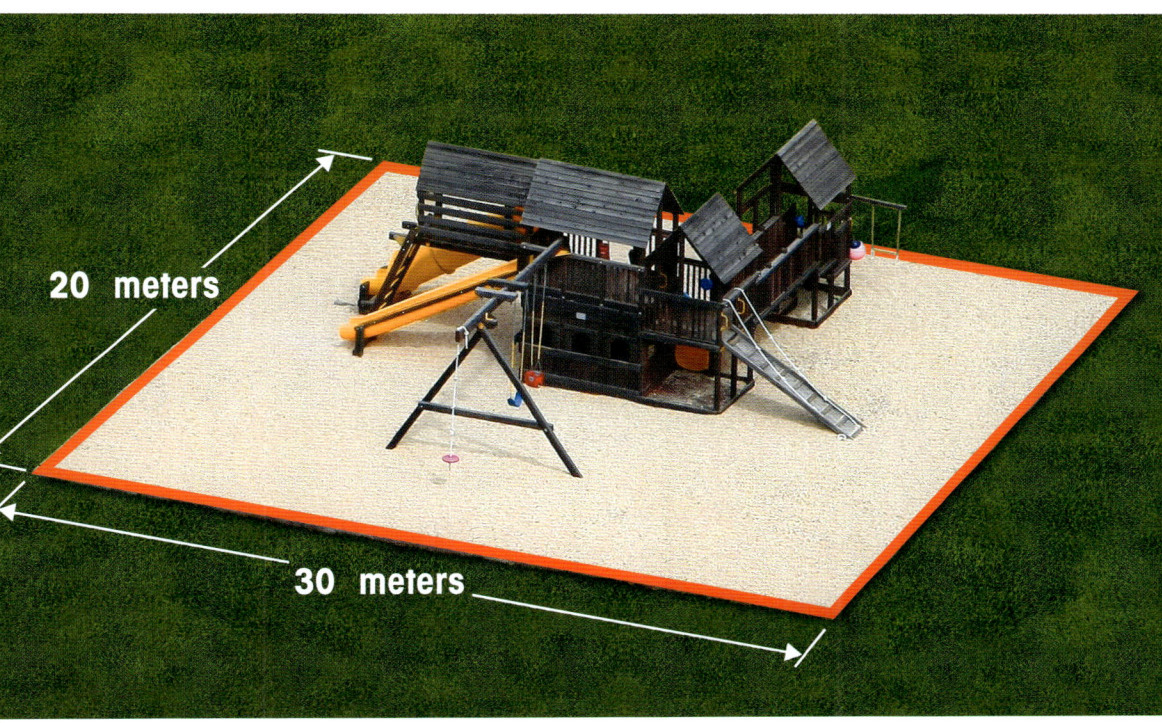

20 meters

30 meters

$A = l \times w$
$A = 30 \times 20$
$A = 600$ square meters

▲ The area is six hundred square meters.

Summary

Many shapes in the park have area. We measure to find the area. We use square units to show area.

Answer: 8 x 4 = 32; 32 square feet

FINDING AREA

Measurement	Formula	Solution
length = 4 centimeters width = 4 centimeters	$A = l \times w$	$A = 4 \times 4$ $A = 16$ square centimeters
length = 22 meters width = 6 meters	$A = l \times w$	$A = 22 \times 6$ $A = 132$ square meters
length = 12 units width = 6 units length = 4 units width = 4 units	$A = l \times w$ Area of **M** + Area of **N**	**M** $A = 12 \times 6$ $A = 72$ square units **N** $A = 4 \times 4$ $A = 16$ square units $A = 72 + 16$ $A = 88$ square units

Talk about the area.

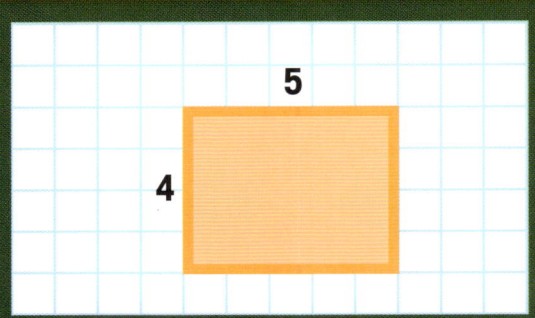

1. The formula for area is ⓘ.
The length is ⓘ units.
The width is ⓘ units.
The area is ⓘ square units.

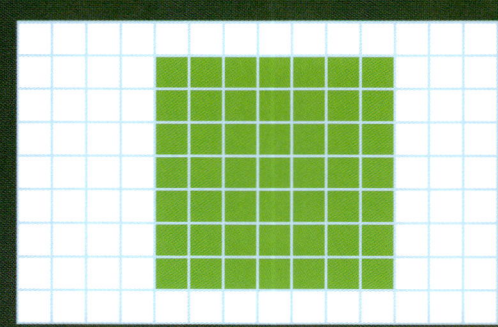

2. The formula for area is ⓘ.
The length is ⓘ units.
The width is ⓘ units.
The area is ⓘ square units.

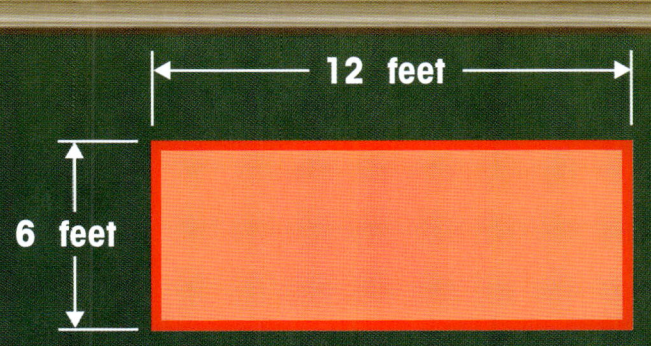

3. The formula for area is (?).
The length is (?) feet.
The width is (?) feet.
The area is (?) square feet.

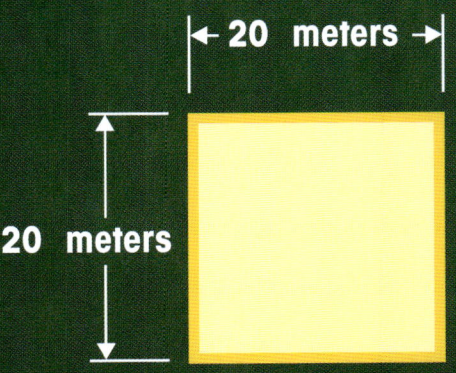

4. The formula for area is (?).

Glossary

A = 12 square units

area (A) the number of square units that cover a shape

*All closed shapes have **area**.*

$$A = l \times w$$

formula a mathematical rule

*This **formula** helps us find the area.*

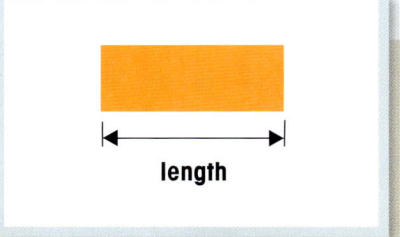

length

length (l) the longer side of a rectangle

*First, we measure the **length**.*

rectangle a quadrilateral with four right angles

*We find the area of the **rectangle**.*

ruler a tool people use to measure

*We use a **ruler**.*

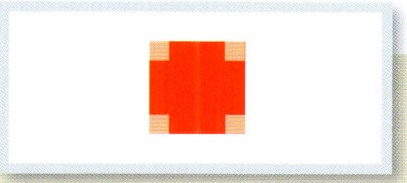

square a polygon with four equal sides and four right angles

*This **square** has an area.*

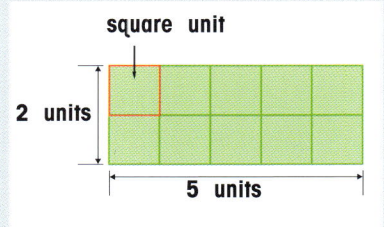

square units units that we use to measure area

*We use **square units** to measure area.*

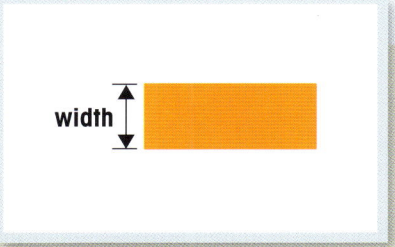

width (w) the shorter side of a rectangle

*Then we measure the **width**.*

Index

After you read . . .

Area at the Park

Materials:

- **paper**
- **pencil**
- **scissors**
- **ruler**

Instructions:

1. Work with a partner.

2. Measure nine square units.

3. Cut out the square units.

4. Put the square units together. Form different shapes.

5. How many shapes can you make? What is the area of each shape?